Cooling Water Treatment Principles & Practices

Summary of Charts and Notes for Field Use

Chemical Publishing Co., Inc., 743 Western Avenue, Gloucester, MA 01930

© 1999, 2010 by Chemical Publishing Co., Inc. & Colin Frayne
All rights reserved. 1999
This book is protected by copyright. No part of it may be reproduced, stored in a retrieval system or transmitted in any form or by any means; electronic, mechanical, photocopying, recording or otherwise, without the prior written permission of the publisher.

ISBN 978-0-8206-0003-1

Made in the United States of America

www.chemical-publishing.com
www.chemicalpublishing.net

Cooling Water Treatment Principles & Practices

Summary of Charts and Notes for Field Use

Colin Frayne

Chemical Publishing Co., Inc

ABOUT THE AUTHOR

Colin Frayne, LRIC, MCIWEM, MICorr. (U.K.) is an international water treatment consultant and small business owner. He has more than 30 years of experience in the practice of industrial chemistry and industrial water systems management, and has worked and lectured in over 40 countries. During those years he has also lived on four continents, with his family, while being variously employed in Q.C. and R & D laboratories, in technical sales, sales management, marketing, training, international business development, import/export, and general management. He graduated in analytical chemistry from North London Polytechnic (now the University of North London), in the United Kingdom, and later obtained various business diplomas from colleges in the U.K. and South Africa, including Wits Business School in Johannesburg. Mr. Frayne is British, but has resided in the United States for several years, with his wife and two daughters. In early 1999 he relocated from Georgia to New York City to join the Metro Group, Inc., an environmental services and water treatment company, in a senior executive capacity.

NOTES

NOTES

NOTES

NOTES

NOTES

NOTES

NOTES

NOTES

NOTES

SUMMARY OF CHARTS AND NOTES FOR FIELD USE

Typically, for refrigeration using **cooling towers**, the following may apply:

$CR = 2.5$ U.S. gpm at 65°F (18.3°C) cold water per ton refrigeration, or 4.0 U.S. gpm at 80°F (26.7°C) cold water per ton, for halocarbon refrigerants
$CR = $ Typically, 60% higher than above for ammonia systems
$\Delta T = 6$ to 15°F (3.3 to 8.3°C)

Typically, for refrigeration using **evaporative condensers**, the following may apply:

$E = 5\%$ of CR
$E = 3.0$ U.S. gpm per 100 tons refrigeration

Typically, for **evaporative cooling towers**, the following may apply:

$E = 1.0\% \times CR$ per 10°F (5.6°C) ΔT (assumes all cooling by evaporation)
$E = 0.9\% \times CR$ per 10°F ΔT (allowing for some sensible-heat transfer)
$E = $ tends to $0.8\% \times CR$ per 10°F ΔT (in areas of high humidity)
$E = $ tends to $1.2\% \times CR$ per 10°F ΔT (in areas of low humidity)

Typically, drift (D) losses are as follows:

$D = 0.3$ to $1.0\% \times CR$ (for natural draft towers)
$D = 0.01$ to $0.3\% \times CR$ (for mechanical draft towers)

Typically, circulation rate (CR) is only 80 to 90% of nameplate/design data

Typically, the volume (V) of system is basin capacity plus 20 to 30%

Bleed: $B = (E/COC - 1) - D$
or: $B = MU - (E + D)$
or: $B = (MU/COC) - D$

Cycles of concentration: COC = Salts in recirculating water/Salts in makeup water

Makeup: $\quad MU = E + B + D$
or: $\quad MU = E \times (COC/COC\ 1)$
Half-life (holding time index): $\quad HL = 0.693 \times V/B$ hr
or: $\quad HL = 2.303 \times (V/B) \times \log_{10} C_i/C_f$

where:

C_i = initial concentration of chemical additive
C_f = final concentration of chemical additive

Conductive heat flow, or *duty*:

Conductive heat flow (Q), or **duty**, of a heat exchanger can be calculated from the following equations (where process temperature changes but does not change state):

$$Q = U \times A \times \Delta T \text{(in Btu/hr)}$$

where: Q = Total conductive heat flow (Btu/hr)
A = Area of the heat-transfer surface (sq ft)
ΔT = LMTD. Differential temperature of two conducting surfaces (in °F)
U = Heat-transfer coefficient (Btu/hr/sq ft/°F)

or: $Q = PF \times SH \times \Delta T$ (in Btu/hr)

where: PF = Process flow, lb/hr
SH = Specific heat of process
ΔT = LMTD. Differential temperature of two conducting surfaces (in °F)

Note that ΔT is approximately equivalent to LMTD (process temperature **in**—process temperature **out**, °F)

The water flow (F_w) through a heat exchanger is (approximately):

$F_w = Q/(t_o - t_i) \cdot (60 \text{ min in hr}) \cdot (8.34 \text{ lb in gal})$ gpm

where: t_o = cooling water temperature **out**
t_i = cooling water temperature **in**

The practical heat-transfer coefficient = ($U_{\text{practical}}$):

Thermal conductance is the reciprocal of resistance (**R**) to heat flow:

$$U = 1/R$$

Summary of Charts and Notes for Field Use

Typical minimum standard specifications for disposal to a water course:

pH:	in the range 6 to 9
BOD_5:	not to exceed 30 ppm
SS:	not to exceed 20 ppm
Heavy metals:	not to exceed 1 ppm
Temperature:	maximum of $25°C$

General industrial and cooling water for makeup purposes commonly contains

- calcium bicarbonate: $Ca(HCO_3)_2$
- calcium chloride: $CaCl_2$
- calcium sulfate: $CaSO_4$
- ferrous bicarbonate: $Fe(HCO_3)_2$
- magnesium bicarbonate: $Mg(HCO_3)_2$
- magnesium chloride: $MgCl_2$
- magnesium sulfate: $MgSO_4$
- manganous bicarbonate: $Mn(HCO_3)_2$
- silica: SiO_2
- silicic acid: H_2SiO_3
- sodium bicarbonate: $NaHCO_3$
- sodium chloride: $NaCl$
- sodium silicate: Na_2SiO_3
- sodium sulfate: Na_2SO_4

Inorganic coagulants

- **Aluminum sulfate**: $Al_2(SO_4)_3 \cdot \times H_2O$
- **Iron salts: ferric chloride**, $FeCl_3 \cdot 6H_2O$, **ferrous sulfate** or **copperas** $(FeSO_4) \cdot 7H_2O$, and **ferric sulfate**, $Fe_2(SO_4)_3 \cdot 3H_2O$
- **Polyaluminum chloride (PAC)**, $Al_n(OH)_m Cl_{3n\ m}$
- **Aluminum chlorhydrate (ACH)**: $Al_2(OH)_5Cl$

Organic coagulants and flocculants

Coagulants: lower MW polymers. Flocculants: higher MW polymers.

- **Acrylamide/acrylate copolymers**: Anionic flocculants
- **Acrylamide/amine copolymers** (+ amine deriv.): Anionic flocculants
- **Mannich polymers**: Cationic, solution polymer flocculants
- **Polyamine/DADMAC**: Cationic coagulants
- **Polyamine/EPIDMA**: Cationic coagulants
- **Polyacrylamide**: Nonionic flocculants
- **Alum + DADMAC or EPIDMA, and PAC + DADMAC or EPIDMA**

Manganese greensand filters

When using manganese greensand for the first time, it should be double backwashed and regenerated with 2 to 4 ounces $KMnO_4$ per cu ft of media.

- Capacity for Fe is 10,000 gal per regeneration at 1 ppm Fe/cu ft media. For Fe/Mn it is 7000 gallons per regeneration, and for H_2S it is 3000 gallons per regeneration at 1 ppm H_2S.
- Maximum practical limits are 15 ppm Fe or Mn and 5 ppm H_2S. Raw water pH is 6.2 to 8.5 and temperature ideally is less than 80°F (26.7°C).
- Size the filter for 30 in. bed depth and gravel under-bed, and a service flow rate of 2 to 5 gpm/sq ft. Backwash typically every 24 hours for 10 minutes at 12 to 14 gpm/sq ft to achieve 35 to 40% bed expansion.
- Limit the pressure drop to 8 to 10 psi maximum to prevent Fe leakage.
- Continuous regeneration is recommended, adding $KMnO_4$ to achieve slight pink color at filter inlet. Also, prefeed Cl_2 10 to 20 seconds upstream of $KMnO_4$ to achieve residual in filter effluent.

Sand filters

Filtration rates are 4 to 6 gpm/sq ft of media surface area, with backwash rates of 10 to 12 gpm/sq ft for a period of 5 to 8 minutes. When the filter is used for removing suspended solids from recirculating cooling water, the filtration rate may be as high as 10 to 15 gpm/sq ft and the backwash rate 15 to 20 gpm/sq ft.

Dual or triple media filters

Flow rates are similar to those for sand filters but higher quality filtered water is generally obtained. Backwash rates need to be lower where anthracite is used to avoid the risk of washing the media out of the filter. Bed expansion is therefore higher at 50 to 100%, depending on grain size and flow rate.

Media for multimedia filters

Media	Sp. Gr.	Grain Size (mm)	Bed Depth (in.)
Anthracite	1.4–1.6	0.80–1.60	18–24
Sand	2.65	0.35–0.80	9–12
Garnet	4.2	0.20–0.40	3–4

Summary of Charts and Notes for Field Use

Water softeners

- Resin type is strong acid cation in sodium form, fully hydrated; typically 53 lb/cu ft at 45% moisture content (0.85 kg/l).
- For continuous softening service it is advisable to install two or more softeners, in parallel, on a duty/standby basis.
- The minimum water pressure required is typically 20 psi.
- Total resin volume design requirement is sufficient resin for 8 hours minimum continuous operation; 12 to 16 hours per tank is fine.
- Typical resin-exchange capacity is from 20,000 grains as $CaCO_3$/cu ft resin (45.9 grams hardness/l resin) at 6 lb NaCl/cu ft resin (0.1 kg/l) to 30,000 grains at 15 lb NaCl/cu ft.
- Typical regenerant concentration is 10 to 20% NaCl; maximum is 36%.
- Resin bed depth is 24 in. minimum, 72 in. maximum (0.61 to 1.83 m).
- Bed expansion (freeboard) is 50% minimum (thus the resin tank needs to be at least double the volume of the resin requirement).
- Service flow rate is 15 to 30 bed volumes (BV)/hour, or 2 to 4 gpm/cu ft resin (0.27 to 0.40 lpm/l resin). Linear flow at 4 to 10 gpm/ft^2.
- Regeneration: Backwash rate is 5 to 6 gpm/cu ft for 10 min (0.69 to 0.80 lpm/l resin).
- Regeneration: Typically, brine injection rate is 1 gpm NaCl soln/cu ft resin (0.13 lpm/l resin) and takes 25 to 30 min. Check NaCl strength. It may take 4 to 16 hr to achieve suitable strength.
- Regeneration: Slow rinse is 1 gpm water/cu ft resin for 15 min (0.13 lpm/l resin).
- Regeneration: Fast rinse is 1.5 gpm water/cu ft resin for 5 min (0.2 lpm/l resin.).

Concentrations of sulfuric acid commonly available

H_2SO_4	Sulfuric Acid			$CaCO_3$ Equivalent to 1 Lb of Acid	
Conc. (%)	Designation	S.G at 60°F	G/L	Lbs	Grains
98.0	98%	1.841	1804	1.0000	7000
93.2	66°Be	1.835	1710	0.9509	6657
77.7	60°Be	1.706	1325	0.7926	5548

- **Corrosion** is an *electrochemical* process whereby the oxidation of metal(s) or alloys to their (lower energy state) oxides or cations takes place, resulting in loss of mechanical or structural strength and metal wastage. Corrosion takes many forms and includes **biocorrosion**, which is corrosion taking place at the water–metal interface of a *biofilm*.

- **Fouling** produces dirty and inefficient cooling systems, impeding the flow of cooling water. It involves the *physical* adherence to surfaces and mutual entanglement of insoluble salts, corrosion products, oils, fats, and other process contaminants, air-blown debris, and the like. Where fouling involves *biomass* (bulky matrixes of microbiological origin), it is often termed **biofouling** and involves the formation of *biofilm*.

- **Deposition** involves the formation and precipitation of crystalline scales and the throwing down, within critical parts of the cooling system, of silt, sand, muds, and sediments; all these *deposit* components have the effect of reducing the rate of heat transfer. Corrosion can also occur under deposits.

Anodic reactions:
1. $Fe \rightarrow Fe^{2+} + 2e$
2. $Fe^{2+} + 2OH \rightarrow Fe(OH)_2$
3. $4Fe(OH)_2 + O_2 + H_2O \rightarrow 4Fe(OH)_3$

Cathodic reactions:
1. $\frac{1}{2}O_2 + H_2O + 2e \rightarrow 2OH$
2. $2H^+ + 2e \rightarrow H_2 \uparrow$

$$H_2O \Leftrightarrow OH^- + H^+$$

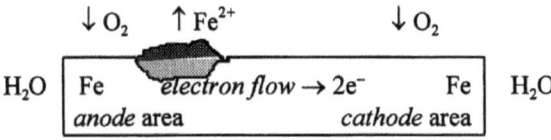

$\downarrow O_2 \quad \uparrow Fe^{2+} \quad\quad\quad \downarrow O_2$

H_2O | Fe electron flow → $2e^-$ Fe | H_2O
anode area   cathode area

Common types of corrosion

- **Oxygen corrosion**
- **Concentration cell corrosion** (*crevice corrosion, under-deposit corrosion, tuberculation, pitting corrosion*)
- **Galvanic corrosion**
- **White rust**
- **Biocorrosion** (from *sulfur bacteria and iron bacteria*)

Summary of Charts and Notes for Field Use

Simple Corrosion Vulnerability Summary

Metal	Type of Corrosion Risk in Cooling Water
Steel	Stagnant water tends to induce local attack
	Fouling induces concentration cell type corrosion
	High Cl⁻ and SO_4^{2-} tend to increase risks of pitting
	O_2-saturated water may induce general corrosion
	Fouling and SO_4^{2-} induce corrosion from sulfate reducers
	Coupling with noble metal causes galvanic corrosion
	pH 4.0 induces rapid acid-induced wastage and pitting
Iron	Presence of some O_2, plus deposits and low flow, stimulates tuberculation
	Soft water, low pH, H_2S can stimulate graphitization
SS	Vulnerable to depassivation from high Cl⁻ and SO_4^{2-}
	Residual or applied stress with Cl⁻ induces SCC
Brass	Ammonia produces cracking and wastage
	High velocity can induce erosion/corrosion (>6 fps)
	Uninhibited brass suffers dezincification
Zinc	High alkalinity causes white rust
Al	pH over 9.0 with OH⁻ causes rapid corrosion

Starting point for specifying and quantifying corrosion rates

Interpretation of Corrosion Rates in Open Cooling Systems		
Description	Rate (mpy)	Comment[a]
Negligible	<2.0	Excellent. Very difficult to achieve
Mild	2.0–4.0	Good. Acceptable for most situations
Moderate	4.0–6.0	OK. Probably adequate protection
High	6.0–8.0	Poor. Doubtful protection
Severe	>10.0	Very poor. Probably unacceptable protection
For copper and brasses, the same comments apply if mpy is 10% of above rates.		

[a]For closed systems, same comments apply if steel mpy is 50% of above rates.

> NOTE: The general corrosion rate of steel in an untreated cooling water system is typically 35 to 40 mpy, so a good chemical treatment program can be expected to reduce this general rate of corrosion by some 90 to 95%.

Calcium sulfate scale

A useful rule for calcium sulfate product concentration is:

- ppm $[Ca^{2+}]$ × ppm $[SO_4^{2-}]$ = 500,000 maximum

Also, there is a significant risk of calcium sulfate deposition occurring in a cooling system where:

- Calcium (as $CaCO_3$) exceeds 1200 ppm at pH 7.0 to 8.0
- Calcium (as $CaCO_3$) exceeds 700 ppm and Cl exceeds 4000 to 5000 ppm

Silica scale

- The practical solubility limit is around *120 ppm at pH 8.0*, rising to *180 ppm at pH 9.5*.

A useful rule for prevention of magnesium silicate deposition is:

Without silica deposit control agent:
- ppm Mg (as $CaCO_3$) × ppm = 20,000 maximum

With silica deposit control agent:
- ppm Mg (as $CaCO_3$) × ppm SiO_2 = 35,000 maximum

Commonly found mineral components of deposits and foulants

Salt/Oxide	Mineral	Formula
Calcium carbonate	Aragonite	$CaCO_3$
	Calcite	$CaCO_3$
Calcium phosphate	Hydroxyapatite	$Ca_{10}(OH)_2(PO_4)_6$
	Tricalcium phosphate	$Ca_3(PO_4)_2$
Calcium sulfate	Anhydrite	$CaSO_4$
	Gypsum	$CaSO_4 \cdot 2H_2O$
	Hemihydrate	$CaSO_4 \cdot \frac{1}{2}H_2O$
Copper oxide	Cuprite	Cu_2O
Iron carbonate	Siderite	$FeCO_3$
Iron oxide	Haematite	Fe_2O_3
	Alpha iron oxide (paramagnetic)	αFe_2O_3
	Gamma iron oxide (ferromagnetic)	γFe_2O_3
	Lepidocrocite	$Fe_2O_3 \cdot H_2O$
	Magnetite	$Fe_3O_4 \cdot H_2O$
Iron sulfide	Troilite	FeS
Magnesium hydroxide	Brucite	$Mg(OH)_4$
Magnesium phosphate	Magnesium hydroxyphosphate	$3Mg_3(PO_4)_2 \cdot Mg(OH)_2$
Magnesium silicate	Serpentine	$3MgO \cdot 2SiO_2 \cdot 2H_2O$
Silica	Crystoballite	SiO_2
Zinc carbonate	White rust	$ZnCO_3$

Summary of Charts and Notes for Field Use

Saturation Indices

$$pH_s = 12.3 \quad (\log_{10} Ca + \log_{10} TA + 0.025 Temp \quad 0.011 TDS^{1/2})$$
$$LSI = pH_{actual} \quad pH_s$$
$$SI = 2pH_s \quad pH_{actual}$$
$$PSI = 2(pH_s) \quad (pH_{eq}),$$

where $pH_{eq} = (1.465 \log_{10} TA) + 4.5$

Equation for solving pH_s

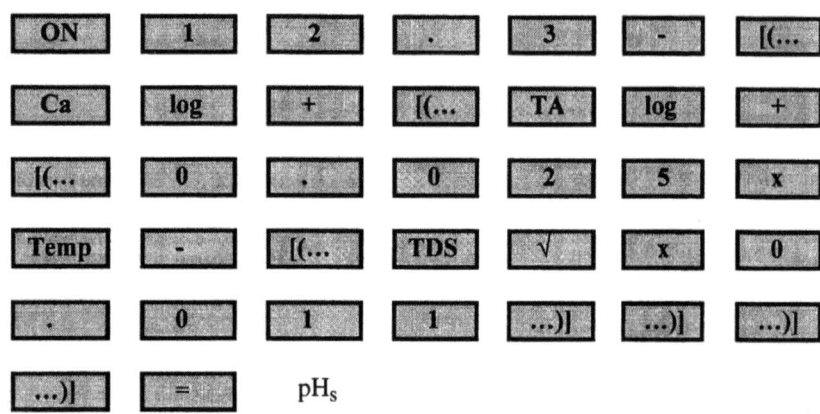

pH_s

BACTERIA

- **Slime formers**: Aerobic, capsulated, gram-negative bacilli, ***Pseudomonas* sp.** (such as *Pseudomonas aeruginosa*), ***Aerobacter* sp.** (also known as *Enterobacter* or *Klebsiella*), ***Bacillus* sp.** (such as *Bacillus subtilis* and *Bacillus cereus*), ***Flavobacterium* sp.**, ***Proteus vulgaris***, ***Serratia* sp.**, and ***Alcaligenes* sp.**
- **Iron bacteria**: Filamentous, aerobic bacteria. Examples are ***Crenothrix polyspora*** and ***Gallionella ferrugine***. Also ***Leptothrix ochracea***, ***Siderocapsa*** and ***Ferrobacillus* sp.**, and ***Thiobacillus ferroxidans***.
- **Sulfur bacteria**: ***Thiobacillus thiooxidans*** is an aerobic, acid- and corrosion-producing, sulfur bacterium. ***Thiothrix* sp.** are troublesome aerobic slime formers. The most prolific **sulfate-reducing bacteria** (*SRB*) is the anaerobe ***Desulfovibrio desulfuricans***. Other sulfur anaerobes include ***Beggiatoa* sp.** and ***Clostridium nigrificans***.
- **Spore formers**: Include ***Bacillus subtilis*** and ***Bacillus mycoides***.
- **Nitrogen convertors**: ***Nitrobacter***, ***Nitromonas***, and ***Nitrococcus* sp.**

- **Pathogenic organisms**: *Legionella* sp., *Escherichia coli*, and *Proteus vulgaris*.

Fungi

- **White and brown rot producers (Basidiomycetes)**: *Poria nigrescans* and *Peniophora mollis* produce white rot. Brown rot is produced by *Poria oleraceae* and *Poria monticola*.
- **Soft rot producers (Ascomycetes and Fungi imperfecti)**: Various fungal organisms are responsible for soft rot; they include, *Bispora*, *Coniothyrium*, and *Stemphylium* sp.

Zooplankton

- **Molluscs**: The *freshwater mussel Dressenia polymorpha*.
- **Marine crustacea**: Includes *acorn* (or *rock*) *barnacle* of group **Cirripedia** and *crabs*. Also the marine **Molluscs**, including **Mytillidae** *mussels*.
- **Protozoa**: *Ameoba proteus*. Also **Paramecium sp**.

Phytoplankton (Algae)

- **Cyanophyta** (*blue-green*)
- **Chlorophyta** (*green*)
- **Rhodophyta** (*red*)
- **Phaeophyta** (*brown*)
- **Chrysophyta** (*yellow*)

Chemical Programs Contain

- **Deposit control agent(s)** or **antiscalent(s)**
- **Dispersant(s)** or **antifoulant(s)**
- **Corrosion inhibitor(s)** for carbon steel
- **Corrosion inhibitor(s)** for copper
- **Microbiocides**
- **Stabilizers: Antifoulants** (*metal surface cleaners*)
- **Secondary inhibitors** to reduce certain side effects

Summary of Charts and Notes for Field Use

Inhibitor Components

- **Amines**
- **Azoles**
- **Chromates**
- **Molybdate**
- **Nitrite**
- **Phosphates**
- **Silicates**
- **Zinc**
- **Polyacrylates**
- **Phosphonates**
- **Phosphinocarboxylic acids**
- **PMA and derivatives**
- **Copolymers**
- **Terpolymers**

Comparisons of Some Solid Bromine/Chloride Products

No.[a]	Chemistry	Form	Size (in)	% Active	% Wt Br	% Wt Cl	Sol. g/100 ml
1	BCDMH	stick	3 × 3/4	96.0	31.8	14.1	0.15
2	BCDMH	tablet	3	96.0	31.8	14.1	0.15
3	BCDMH	granule		96.0	31.8	14.1	0.15
4	60% BCDMH 27% DCDMH 11% DCEMH	briquette	1.65 × 0.9	98.0	19.6	20.0	0.54
5	BCDMH	tablet	1	98.0	30.6	13.6	0.20
6	BCDMH	granule		98.0	30.6	13.6	0.20
7	93% TCCA 7% NaBr	tablet	2.8	93.4	5.1	39.8	1.20
8	93% DCCA 7% NaBr	granule		89.4	4.9	26.8	26.0

NOTE: Products No. 2, 3, 5, and 6 were recently upgraded from 92.5–93.5% active.
TCCA = Trichloroisocyanurature, DCCA = Dichloroisocyanurate

Availability of undissociated hypochlorous and hypobromous acids in cooling water at various levels of pH

pH	% HOCl	% HOBr	% HOI
6.5	95	100	
7.0	90	100	
7.5	50	94	84
8.0	24	83	
8.5	9	60	
9.0	3	33	97
9.5	0	11	

Suggested biocide selection for comfort cooling systems

pH range	Algae and Algal Biomass	Gram-negatives and Biofilms	SRBs and Heavy Slimes
below 7.5	Guanides or sulfone. All ± isocyanurate	Alkyl sulfonate or DBNPA. All ± isocyanurate	Thione or THPS. Or Glut. + BCP. All ± isocyanurate
7.5–8.5	Guanides or polyquat. All ± Towerbrom	TTPC or DTEA. Or Glut. + BCP. All ± Towerbrom	Thione or THPS. Or Glut. + BCP. All ± Towerbrom
above 8.5	Guanides or polyquat. All ± BCDMH	TTPC or DTEA. Or Glut. + BCP. All ± BCDMH	Trisnitro or THPS Or Glut. + BCP. All ± BCDMH

Possible biocide selection for large industrial cooling systems

pH range	Algae and Algal Biomass	Gram-negatives and Biofilms	SRBs and Heavy Slimes
below 7.5	Chlorophenol. Or MBT + TCMTB + chlorine	Alkyl sulfonate. Or MBT + TCMTB All ± chlorine	Alkyl sulfonate. Or MBT + TCMTB All ± Chlorine
7.5–8.5	Chlorophenol. Or TBZ ± isothiazoline. ± bromine	BHAP or DTEA. Or isothiazoline ± polyquat. All ± bromine	Trisnitro. Or Isothiazoline ± polyquat All ± bromine
above 8.5	Chlorophenol. Or TBZ ± isothiazoline. ± bromine	BHAP or DTEA. Or isothiazoline ± polyquat. All ± bromine	Trisnitro. Or isothiazoline ± polyquat All ± bromine

Summary of Charts and Notes for Field Use

Deposition:
of scales, silt, muds & sediments

Corrosion:
inc. general wastage, pitting biocorrosion & loss of metal strength

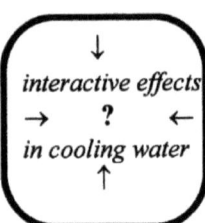
interactive effects
→ ? ←
in cooling water

Fouling:
by salts, oils, corrosion prods. air-blown debris & biofouling

Microbiological growth:
inc. planktonic & sessile bacteria. And formation of biofilms, slimes & biomass

Inhibitor Classification:	Zinc Chromate				
Basic premise:	Destroy alkalinity with acid. No $CaCO_3$ scale! Then provide excellent corrosion inhibition with synergistic Zn/CrO_4				
Application:	Only for those plants with CrO_4 discharge consents or "zero discharge," or with facilities for treating cooling tower bleed.				
Notes:	With lower acid usage, or higher level of Ca in makeup: increase polymer/phosphonate.				
Operating ranges:	CrO_4	PO_4	DCA	pH	LSI
Std chromate	±15	±3	±4	6.0–7.0	0.0 max
Low chromate	±10	±5	±6	7.0–7.5	+0.5 max
Very low chromate	±5	±7	±8	7.5–8.0	+1.0 max

Inhibitor Classification	Stabilized Phosphate (SP)			
Basic premise:	Use acid. ± Neutral LSI. Add O + P-PO_4 for corrosion protection. Avoid risk of $Ca_3(PO_4)_2$ deposition by use of "stabilizer" dispersant.			
Application:	Larger systems! Needs good chemistry skills			
Notes:	Available with/without Zn. As Ca hardness increases (up to 1200 ppm), decrease PO_4 and increase stabilizer (8–25 ppm *actives*).			
Operating ranges:	O:P ratio	PO_4	pH	LSI
SP for lower hardness	2.0 : 1.0	±16	6.8–7.4	+0.5 max
SP for high hardness	3.0 : 1.0	±12	7.4–7.8	+1.0 max

Inhibitor Classification:	Alkaline Phosphate				
Basic premise:	No acid used! Maintain lower O + P PO_4 reserve. Original objective in early programs was to balance corrosion/deposition risk. Now, the increased risk of $CaCO_3$ deposition is controlled with improved organic DCAs.				
Application:	*Large and small C/S. Simple formulations have limited tolerance for high temp/stress/pH.*				
Notes:	Limit ±600/700 ppm Ca hardness as $CaCO_3$				
Operating ranges:	Zn	PO_4	DCA	pH	LSI
Alkaline Zinc	±3	±7	±5	7.0–8.0	+1.5 max
Alkaline Phosphate	Nil	±7	±8	8.0–9.0	+2.5 max
Zn Phosphate Organic	±3	±7	±11	7.5–9.0	+2.5 max

Inhibitor Classification:	Zinc Polymer Phosphonate (ZPP) (Zn Organic)				
Basic premise:	No acid! Phosphonate replaces PO_4. Zn can be replaced with Mn. Organics have become equal status corrosion inhibitors (e.g., HPCA/heptonates). DCAs control deposition.				
Application:	*Large and small C/S. Simple formulations have limited tolerance for high temp/stress/pH.*				
Notes:	Limit ±800 ppm Ca hardness as $CaCO_3$				
Operating ranges:	Zn	Org C.I.	DCA	pH	LSI
ZPP for simple C/S	±3	Nil	±10	7.5–8.5	+1.5 max
ZPP for low Ca MU	±3	±6	±7	7.0–8.5	+1.0–2.0
ZPP for high Ca MU	±2	±4	±10	8.0–9.0	+2.5 max

Inhibitor Classification:	All Organic Programs (AOP)				
Basic premise:	No acid, metals, or PO_4! Organics provide all inhibitor functions, but operate at higher alkalinity to minimize corrosion potential				
Application:	*Large and small C/S. Suitable for wide range of MU water chemistries and low-high stress*				
Notes:	Limit: ±1000 ppm T. Alk. Prevention of pitting corrosion needs clean metal surfaces!				
Operating ranges:	Total actives		Org. P	pH	LSI
AOP for simple C/S	±12	ppm	ATMP	7.5–9.0	+2.0
AOP for high stress	±16	ppm	PBTC	8.5–10	+3.0

Summary of Charts and Notes for Field Use

Inhibitor Classification:	Organic + Tracer				
Basic premise:	No acid required. Organic inhibitor base, but with low Zn, Mo, PO$_4$, or inorganic mix, which adds capability for tracing, and also aids low-end, pitting corrosion control				
Application:	Large and small C/S. Suitable for wide range of MU water chemistries and low-high stress				
Notes:	For Zn or PO$_4$: Limit to 700 ppm T. Alk. For Mo tracer: Limit to 1000 ppm T. Alk.				
Operating ranges:	Tracer	Org C.I.	DCA	pH	LSI
PO$_4$ OT for lean water	±4	±6	±8	7.5–8.5	+2.0
Mo OT for high stress	±1.5	±2.5	±12	8.0–9.5	+3.0

Inhibitor Classification:	Molybdates				
Basic premise:	No acid required. MoO$_4$ is expensive, but a good replacement corrosion inhibitor for CrO$_4$ and stabilized S/PO$_4$. Costs lowered when synergized with phosphonates, as lower levels needed				
Application:	Small-medium C/S. Good for lean MU water				
Notes:	Keep cycles high for max. MoO$_4$. Keep clean C/S to minimize pitting. Simple accurate test				
Operating ranges:	Mo	Org P	DCA	pH	LSI
High Mo (lower TH)	5–7	2–3	±7	±8.0	+0.5 to +1.5
Low Mo (higher TH)	3–4	3–4	±11	8–9	+1.5 to +2.5

Inhibitor Classification:	Polysilicates				
Basic premise:	"Natural" scale/corrosion inhibitor. Potential for "environmentally friendly" or "green" product marketing.				
Application:	Small-medium C/S. Good for soft and lean MU waters, where high corrosion risk exists				
Notes:	Program needs good silicate dispersant polymer, to avoid risk of silicate scales. Also, provide good passivation and clean system. Allow for existing silicates in water.				
Operating ranges:	MU	T Hard	SiO$_2$	pH	LSI
Silicate/org./Zn + Mo	soft	0–10	±25	6.5–8.0	1 to +1
Silicate/PO$_4$/organic	lean	10–25	±15	7.0–8.5	0 to +1.5
Silicate/nitrite/organic	low	25–75	±12	7.5–9.0	1 to +2

Conductivity (μS/cm) × 0.50 = TDS ppm NaCl

Conductivity (μS/cm) × 0.65 = TDS ppm for a typical water

TDS/conductivity meters can be standardized with a **potassium chloride (KCl)** solution, as per ASTM D-1125-77.

$$0.001 \text{ N KCl} = 147 \text{ μS/cm}$$
$$0.01 \text{ N KCl} = 1409 \text{ μS/cm}$$
$$0.1 \text{ N KCl} = 12850 \text{ μS/cm}$$

The control of Legionellosis

- Identification and assessment of risk
- Preparing a scheme for preventing or minimizing the risk and applying precautions to control the risk
- Allocation of specific responsibilities, training, and other resources
- Setting up and maintaining adequate records and control procedures
- Methods for decontamination and cleaning—"environmental services"

Protocols for cooling system cleaning and disinfection programs

- **Prechlorination stage**: Initially disinfect tower water with 5 ppm free reserve of chlorine maintained for at least one hour at pH 7.0 to 7.6
- **Draining and cleaning stage**: Circulate for 1 to 4 hours, then drain and physically clean system. Biodispersants may be employed.

  Circulate for 1–4 hours, then drain and physically clean system. Biodispersants may be employed.

- **Disinfection, recommissioning, and water treatment reinstatement stage**: The system is refilled and disinfected for the required period, then put back into service. Routine maintenance: water is considered disinfected if 5 ppm free reserve of chlorine is maintained for at least one hour. Decontamination because of positive *Legionella* results: 10 ppm free reserve for at least 1 hour at pH 7.0 to 7.6.

Summary of Charts and Notes for Field Use

$$\text{Corrosion rate, } \mathbf{mpy} = \frac{\text{area factor} \times \mathbf{mg} \text{ weight loss}}{E}$$

given that E = exposure time, in days.

$$\text{Area factor} = \frac{22.3}{DA}$$

D = density of metal, g/cm^3 (carbon steel = 7.85, admiralty brass = 8.2 to 8.5, aluminum = 2.70, copper = 8.9)
A = area of coupon exposure, in.^{2}.

$$\text{Corrosion rate, } \mathbf{mdd} = \frac{\text{mpy} \times D}{1.437}$$

$$\text{Pitting rate, } \mathbf{PR} = \frac{\text{max. pit depth} \times 365}{E}$$

Cooling water control parameters
These notes are provided as an outline guide only. See the relevant section for more details. Precise control parameters are site-specific.

- **Alkalinity**: *Non-acid-treated cooling system* = ±600 ppm Total Alk. CaCO$_3$ max. *With high alk. water plus good DCA*, perhaps 800–1000 ppm, Total Alk. *For acid treated* = typically 30–80 ppm, Total Alk. CaCO$_3$
- **Aluminum**: Typically limit to ±1.0 ppm total Al maximum.
  Typically limit to ±1.0 ppm total Al maximum.
- **Bacteria levels**: For *industrial and building services*: 1 × 10^5 cfu TAB maximum. For *hospital and health care premises*: 1 × 10^4 cfu TAB maximum.
- **Calcium hardness**: Traditionally = ±800 ppm Ca Hardness as CaCO$_3$.
- **Chloride**: <200 ppm OK. Corrosion risk increases with higher Cl . Keep clean!
- **Copper**: Typically limit to (0.05 ppm) Cu maximum.
- **Cycles of concentration**: Variable! Typical practical = 2.5 × 6.0x
- **Electrical conductivity**: Generally not critical. TDS × 1.54 = µS/cm.
- **Iron**: Typically limit to ±0.5 ppm total.
- **Legionella testing**: *Positive* result below 1 × 10^2: inspect and retest. Positive result above 1 × 10^2: cleaning/disinfection program is necessary.

- **Managenese**: Typically limit to ±0.5 ppm Mn maximum.
- **Molybdate**: Tracer = 1–1.5 ppm. As corrosion inhibitor = >7–10 ppm.
- **Oil leaks**: Free oil should never exceed ±5 ppm.
- **pH**: Variable within range 6.5–9.5. Typical today = pH 7.5–9.0.
- **Phosphate**: If present use PO_4 DCA. Programs vary from 2 ppm to 20 ppm PO_4.
- **Silica**: Practical limit is ±*120 ppm at pH 8.0* rising to *180 ppm at pH 9.5*.
- **Sulfates**: Limit to ppm $[Ca^{2+}]$ × ppm $[SO_4^2\ ]$ = 500,000 maximum.
- **Sulfides**: Typically limit to ±1 ppm S maximum.
- **TDS**: Typically limit to (2500–3500 ppm) TDS maximum.
- **Zinc**: Often added as part of program. Typical = 1.5–3.0 ppm Zn.

Analytical Conversions of Some Common Substances

Substance	Formula	Atomic/ Molecular Wt.	Valency	Gram Equiv. Wt	1 meq/l equals ? ppm
Bicarbonate	HCO_3	61.0	1	61.0	61.0
Calcium	Ca	40.1	2	20.0	20.0
Calcium bicarbonate	$Ca(HCO_3)_2$	162.1	2	81.1	81.1
Calcium phosphate	$Ca_3(PO_4)_2$	310.3	6	51.7	51.1
Calcium sulfate	$CaSO_4$	136.1	2	68.1	68.1
Chloride	Cl	35.5	1	35.5	35.5
Iron (ferrous)	Fe^{2+}	55.8	2	27.9	27.9
Iron (ferric)	Fe^{3+}	55.8	3	18.6	18.6
Ferrous hydroxide	$Fe(OH)_2$	89.9	2	44.9	44.9
Silica	SiO_2	60.1	2	30.0	30.0
Sodium chloride	NaCl	58.5	1	58.5	58.5

Interpretation of P and M Alkalinity Titrations

Ions	$P = 0$	$P < M/2$	$P = M/2$	$P > M/2$	$P = M$
(OH)	0	0	0	2P − M	M
CO_3	0	2P	M	2(M − P)	0
HCO_3	M	M − 2P	0	0	0

For **$CaCO_3$**: (MW = 100): 100 mg/l = 100 g/m^3 = 100 ppm = 2 meq/l = 5.848 grains per U.S. gal (grpg) = 7.00°E = 5.559°D = 10.0°F = N/500 = 0.84 lb per 1000 U.S. gal = 1 lb per 1000 imperial gal.

NOTES

NOTES

NOTES

NOTES

NOTES

NOTES

NOTES

NOTES

NOTES

NOTES

Other publications available by:

COLIN FRAYNE
CSci, CChem, CEnv, CWEM, FRSC, FICorr,
MCIWEM, MWMSoc (UK), CWT (USA)

COOLING WATER TREATMENT:
PRINCIPLES AND PRACTICE

512 pages, Illustrations, 1999, Hardcover
ISBN: 978-0-8206-0370-4

BOILER WATER TREATMENT:
PRINCIPLES AND PRACTICE

400 pages (Vol 1), 548 (Vol 2)
Illustrations, 2002, Hardcover
Vol 1 ISBN: 0-8206-0371-6
Vol 2 ISBN: 0-8206-0400-3

Chemical Publishing Co., Inc.
P.O. Box 5359 • Gloucester, MA 01930
Telephone: 978.525.CHEM (2436)
Fax: 978.525.3452

www.chemical-publishing.com
www.chemicalpublishing.net

Printed by Libri Plureos GmbH in Hamburg, Germany